The County of Edinburgh or Midlothian. Its geology ... Second edition.

Ralph Richardson

The County of Edinburgh or Midlothian. Its geology ... Second edition.
Richardson, Ralph
British Library, Historical Print Editions
British Library
1895
40. pp. viii. 7-40 p. ; 8º.

The BiblioLife Network

This project was made possible in part by the BiblioLife Network (BLN), a project aimed at addressing some of the huge challenges facing book preservationists around the world. The BLN includes libraries, library networks, archives, subject matter experts, online communities and library service providers. We believe every book ever published should be available as a high-quality print reproduction; printed on- demand anywhere in the world. This insures the ongoing accessibility of the content and helps generate sustainable revenue for the libraries and organizations that work to preserve these important materials.

The following book is in the "public domain" and represents an authentic reproduction of the text as printed by the original publisher. While we have attempted to accurately maintain the integrity of the original work, there are sometimes problems with the original book or micro-film from which the books were digitized. This can result in minor errors in reproduction. Possible imperfections include missing and blurred pages, poor pictures, markings and other reproduction issues beyond our control. Because this work is culturally important, we have made it available as part of our commitment to protecting, preserving, and promoting the world's literature.

GUIDE TO FOLD-OUTS, MAPS and OVERSIZED IMAGES

In an online database, page images do not need to conform to the size restrictions found in a printed book. When converting these images back into a printed bound book, the page sizes are standardized in ways that maintain the detail of the original. For large images, such as fold-out maps, the original page image is split into two or more pages.

Guidelines used to determine the split of oversize pages:

• Some images are split vertically; large images require vertical and horizontal splits.
• For horizontal splits, the content is split left to right.
• For vertical splits, the content is split from top to bottom.
• For both vertical and horizontal splits, the image is processed from top left to bottom right.

The

County of Edinburgh

or Midlothian

Its Geology, Agriculture, & Meteorology

WITH

An Agricultural Map of the County, &c., &c.

BY

RALPH RICHARDSON, F.R.S.E.

COMMISSARY CLERK OF THE COUNTY OF EDINBURGH; HONORARY SECRETARY OF THE
ROYAL SCOTTISH GEOGRAPHICAL SOCIETY, ETC.

SECOND EDITION

With additions relating to the Mineral Oil Industry of the County,
Population and Personalty,
and Meteorological and Agricultural Statistics.

EDINBURGH
WILLIAM F. CLAY, 18 TEVIOT PLACE
1895

THE

COUNTY OF EDINBURGH, OR MIDLOTHIAN

ITS

GEOLOGY, AGRICULTURE, AND METEOROLOGY

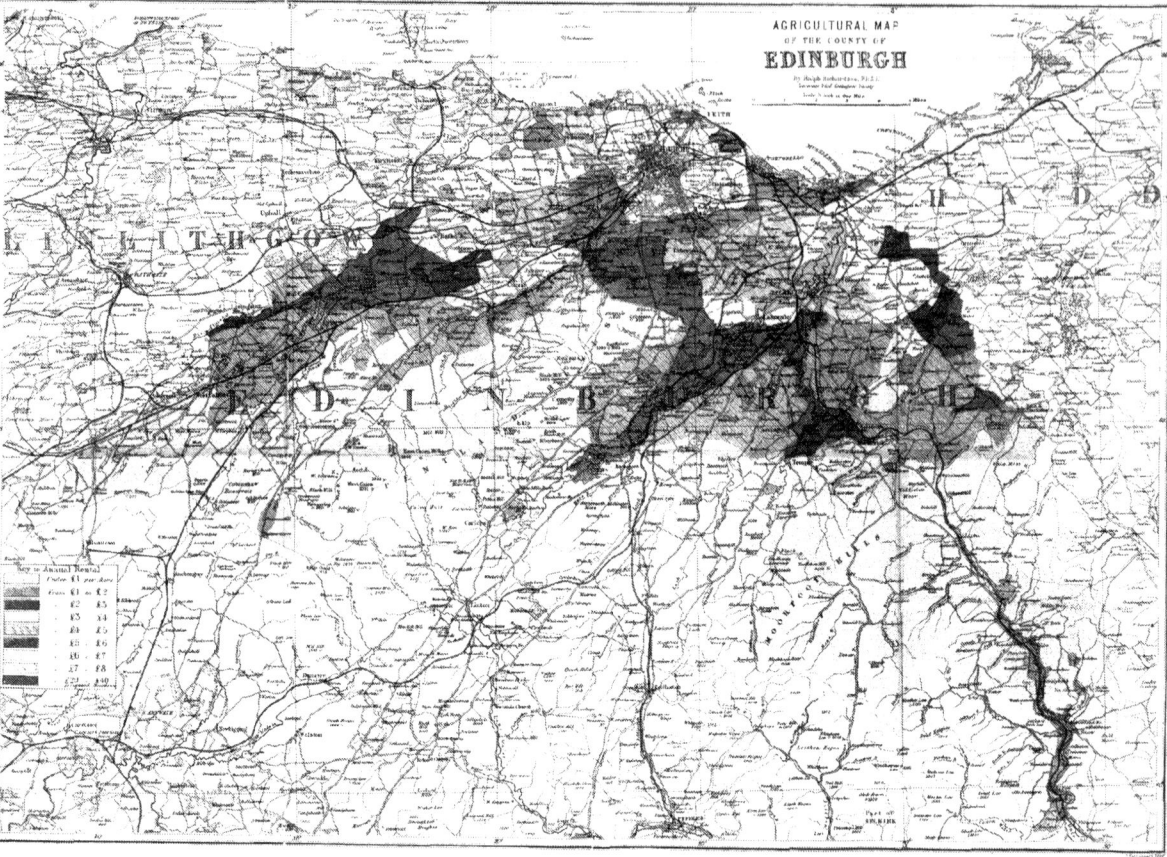

AGRICULTURAL MAP
OF THE COUNTY OF
EDINBURGH

The

County of Edinburgh

or Midlothian

Its Geology, Agriculture, & Meteorology

WITH

An Agricultural Map of the County, &c., &c.

BY

RALPH RICHARDSON, F.R.S.E.

COMMISSARY CLERK OF THE COUNTY OF EDINBURGH; HONORARY SECRETARY OF THE
ROYAL SCOTTISH GEOGRAPHICAL SOCIETY, ETC.

SECOND EDITION

With additions relating to the Mineral Oil Industry of the County,
Population and Personalty,
and Meteorological and Agricultural Statistics.

EDINBURGH
WILLIAM F. CLAY, 18 TEVIOT PLACE
1895

First Edition published 1878

Second Edition published 1895
BY
WILLIAM F. CLAY, 18 TEVIOT PLACE
EDINBURGH

PREFACE TO SECOND EDITION.

IN re-issuing this little book, I find nothing to alter in the elementary descriptions which it contains of the Geology, Agriculture, and Meteorology of Midlothian. With regard to the Agricultural Map, it still remains, I think, the first and only map of the kind yet published in Great Britain. The map was exhibited at the International Exhibition held at Edinburgh in 1886, and I have never heard its general accuracy disputed.

Fully aware, however, of the great agricultural depression which has occurred since the map was published, I submitted it (previous to its present re-issue) to Mr Kennedy, the highest authority in the county on the value of land, and he informs me that he considers that it still fairly represents the average rental of Midlothian.

The long period of thirty-three years elapsed between the publication, in 1859, of the *first* edition, and, in 1892, of the *second* edition, of the Geological Survey Map of the Edinburgh District. Mr H. M. Cadell of Grange (himself one of the authors of the second edition) contributed an interesting paper upon the new geological map to the seventh volume (1894) of the *Transactions of the Edinburgh Geological Society.* From that paper I extract the following remarks :—" The most important additions to the map are to be found in the triangular area between the Firth of Forth and Cobbinshaw Reservoir, at the south-west corner of the sheet, within which the more important fields of *oil shale* are situated. A glance at the West Calder district shows how complicated the structure is in some localities. The rocks are not only very variable in thickness, but are bent about into multitudes of irregular folds, basins, and domes. These are in turn cut up in all directions by faults of varying dimensions, some of which run for miles across the country and extend westwards into the adjoining sheet, where their continuations may be followed into the coalfields of Lanark and Linlithgowshire. . . .

"In the West Calder district, the whole succession from the Burdiehouse or Camps estuarine Limestone may be traced upwards to the Hurlet or Mountain Limestone, and it is within this zone that the majority of *workable oil shales* have hitherto been found. A list of the seams of shale is given on the map, and several descriptions of the section of the oil-shale group of the Calciferous Sandstone series have been published within the last few years from information derived from mining sections. In the Broxburn and West Calder district, the general section of the series, as published in a paper by me in the *Journal of the Iron and Steel Institute* for 1888, is as follows :—

Strata below the Hurlet Limestone, about	400 feet.
Raeburn's Shale, 3 to 4 feet.	
Strata, ...	190 ,,
Mungal's Shale, about 1 foot 9 inches.	
Strata, ...	170 ,,
Coal and Grey Shale, 2 feet.	
Strata, Houston Marl Group, ...	200 ,,
Houston Coal and Shale, 3 to 12 feet.	
Strata, usually laminated Sandstones, ...	240 ,,
Fell's Shale, 3 to 5 feet.	
Strata, Broxburn Marl Group, . . .	80 to 270 ,,
Broxburn Shale, 2½ to 8 feet.	
Strata, including the Binny Sandstones, . . .	450 ,,
Dunnet Shale, 6 to 13 feet.	
Strata, thickness imperfectly known, say . . .	400 ,,
Barrack's Shale, resting at places on the Burdiehouse, Camps, or Queensferry Limestone, thickness very variable.	
Strata, about	780 ,,
Pumpherston Shales.	
Apparent approximate thickness of OIL-SHALE Group, .	3100 feet."

The Mineral Oil Trade of the county, in which so much capital and labour are engaged, has known many vicissitudes. At this moment it seems to be emerging from severe depression, and the hope may be expressed that it will once more prove a successful enterprise.

As pointed out by Mr H. M. Cadell in his paper on "Some Ancient Land-Marks of Midlothian," which appeared in *The Scottish Geographical Magazine* for June 1893, the vicinity of Edinburgh was once studded with Lakes, of which—in sorely shrunken shapes—Duddingston and Lochend lochs are the sole survivors. In addition to these were the lochs of Gogar, Corstorphine, Craigcrook, Holyrood, and the Burgh Loch (the Meadows). A map accompanying the paper indicates their positions, and shows how well watered the neighbourhood of Edinburgh once was.

So far as accumulated wealth is concerned, the County of Edinburgh is one of the wealthiest districts in the world. Some interesting statistical conclusions may be drawn from the following table, showing the

COMPARATIVE DECENNIAL INCREASES OF THE POPULATION OF, AND THE PERSONALTY (exclusive of that under English and Irish Probates and Letters of Administration) given up in, the COMMISSARIOT OF EDINBURGH, which embraces the Parliamentary Burghs of Edinburgh, Leith, Musselburgh, and Portobello, and all Midlothian. (As the Metropolitan Court for Scotland, estates are likewise given up in the Commissary Court at Edinburgh of persons owning Scottish personalty dying domiciled furth of Scotland, or where the Scottish county or place of domicile are uncertain or unknown. Although included in the subjoined table, such personalty does not probably average more than about 15 per cent. of the whole.)

| Census Year. | Population of County. | Personalty given up within Year. | Increase in each 10 Years of | |
			Population per Cent.	Personalty per Cent.
1861	273,869	£1,772,440	...	...
1871	328,335	2,845,043	20	60
1881	388,977	2,978,818	18	
1891	434,276	4,137,721	12	40

As regards her contributions to the Imperial revenue, Edinburgh enjoys a proud pre-eminence. In a speech delivered before the Edinburgh Merchant Company on 30th November 1894, the Earl of Rosebery, K.G., First Lord of the Treasury,

and Lord Lieutenant of the County of Edinburgh, remarked:—
" Edinburgh occupies a very exceptionable position in regard to
the Revenue. I wonder if any of you know it. What relation
do you suppose Edinburgh plays to the other cities of the
United Kingdom in regard to the Revenue? Of those who
pay contributions to the Revenue of the United Kingdom,
Edinburgh stands second in regard to its proportion to the
Revenue Board. London stands first; Edinburgh stands second."
(*Scotsman*, 1st December 1894.)

I am indebted to Mr Peter Loney, George Square, Edinburgh,
for details regarding the Temperature and Rainfall at Edinburgh
from 1877 to 1894 inclusive; The Fiars Prices for Midlothian
from 1877 to 1894 inclusive; and the Average Prices of Wheat,
Barley, Oats, Peas and Beans, and Oatmeal, for ten years from
1877 to 1886 inclusive, and for eight years from 1887 to 1894
inclusive. Mr Loney's statistics will prove interesting not only
to agriculturists and meteorologists, but also to that wider public
which recognises how closely the sciences of Agriculture and
Meteorology are linked together.

R. R.

EDINBURGH, *1st June 1895.*

THE COUNTY OF EDINBURGH,

OR

MID-LOTHIAN:

ITS GEOLOGY, AGRICULTURE, AND METEOROLOGY.

INTRODUCTION.

THE publication of an Agricultural Map of France in the Bulletin of the Geographical Society of Paris for July 1874 has been the origin of the present little treatise. The map referred to was prepared by M. Achille Delesse, Ingenieur en Chef des Mines, and Membre de la Société Centrale d'Agriculture. He describes it in his "Revue de Géologie* pour les années 1874 et 1875" (pp. 185–189), to which a copy of the map is appended. M. Delesse had devoted himself specially to the study of the soil of France, having collected from the various districts of that country upwards of six thousand samples of soils. (Paper by Mr James Melvin, Bonnington, "Trans. Edin. Geol. Soc." vol. iii. part i., 1877.) The author had the pleasure of an interview with M. Delesse in Paris in the spring of 1877, and of seeing his large Agricultural Map of France, of which that above referred to is only the very much reduced copy. In addition to viewing Mid-Lothian agriculturally, the author has thought it might be interesting and advantageous to consider its geological and meteorological features as well. There is a distinct connection between the geology, agriculture, and meteorology of a district. "While the *climate* determines the general character of the vegetable produce, and what kind of plants, under the meteorological conditions, can arrive at perfection, yet the *geological structure*

* Tome xiii. Paris. Librarie F. Savy. 1877.

determines and enables us to judge beforehand, to a certain extent, whether or not any crops shall be able to grow at all, and of the kind of plants suitable to the climate which can be profitably cultivated." (Johnston and Cameron's "Elements of Agricultural Chemistry and Geology," 1877). A few general observations on the county may also be introduced by way of preface.

GENERAL OBSERVATIONS.

The metropolitan county of Scotland extends about 30 miles from west to east, and about 24 miles from north to south. Its area is 367 square miles, or 234,926 imperial acres. Its valued rental for 1877-78 is £578,583 (exclusive of railways); valuation of railways (exclusive of portion situated within burghs), £104,965. The population of the county in 1871 was 328,379 (exclusive of Parliamentary burghs). It contains the following Parliamentary burghs :—EDINBURGH (pop. 1871, 196,979; estimated pop. 1877, 218,729); yearly rent or value 1877-8, £1,538,738). LEITH (pop. 1871, 44,280; estimated pop. 1877, 54,257; yearly rent 1877-8, £327,038). MUSSELBURGH (pop. 1871, 7516; yearly rent 1877-8, £22,663); and PORTOBELLO (pop. 1871, 5481; yearly rent 1877-8, £40,666). The amount of income on which income-tax was paid by residenters in the city and county of Edinburgh in 1877 was £7,193,000. (Mr D. M'Laren, M.P., 24th Sept. 1877.)

As regards the physical features of the county, the most striking is its being bounded on the south by the Pentland and Moorfoot ranges of hills, from which (excepting Gala Water) the streams run northwards to the Firth of Forth, which forms the northern boundary of the county. Of these hills, Carnethy, in the Pentlands, rises to 1890 feet, and Blackhope Scar, in the Moorfoots, to 2136 feet. The city of Edinburgh itself stands high; Princes Street being 209 feet, and the top of Bruntsfield Links 312 feet high. Those familiar objects, the Calton Hill, Castle Rock, and Arthur's Seat, rise respectively 355, 437, and 823 feet above the sea-level.

The county of Edinburgh is divided into 32 parishes, and contained in 1871, 27,856 inhabited houses. According to the Parliamentary Return for 1872-3 of Owners of Lands and Heritages (17 & 18 Vict. c. 91), commonly called the "Domesday Book" of Scotland, the *county* of Edinburgh contained 226,778 acres, of a gross annual value of £581,603, 6s.; there were 696 owners of land of one acre and upwards, and 2541 owners of lands of less than one acre in extent. The former class of owners possessed 226,223 acres, of a gross annual value of £535,200, 1s.; the latter class possessed 555 acres, of a gross annual value of £46,403, 5s. These statistics, from the Parliamentary Return, are exclusive of the municipal burghs of Edinburgh and Leith, the returns for which were as follows :—

EDINBURGH.

		Estimated Acreage.	Gross Annual Value.
Owners of land of one acre and upwards,	240	2558	£252,967
Owners of land of less than one acre, .	11,306	1180	1,041,364
Total, ...	11,546	3738	£1,294,331

LEITH.

		Estimated Acreage.	Gross Annual Value.
Owners of land of one acre and upwards,	127	956	£111,658
Owners of land of less than one acre, .	2062	270	141,446
Total, ...	2189	1226	£253,104

According to the same Parliamentary Return, the following were in 1872-3 the landowners in the county of Edinburgh, who owned 1000 acres or more, or land of the annual value of £1000 or upwards :—

Name of Landowner.	Estimated Acreage of Property.	Gross Annual Value.
Abercorn, Duke of, Duddingston House . .	1,500	£7,400
Aberdour, Lord, Dalmahoy . . .	1,467	5,411
Baird, Sir D., of Newbyth, Bart. . . .	751	3,456
Do. (Mines) . . .	...	400
Baird, Sir J. Gardiner, of Saughtonhall, Bart. .	340	2,450
Borthwick, John, of Crookston . . .	5,239	4,366
Brash, Heirs of Peter, Leith . . .	285	1,237
Do. (Mines) . .	...	500
Brown, H. R. F., of Newhall, Curators of . .	1,635	1,062
Buccleuch, Duke of, Dalkeith Palace . .	3,532	16,216
Do. (Mines) . .	...	1,479
Do. (Granton Harbour)	9	10,601

Name of Landowner.	Estimated Acreage of Property.	Gross Annual Value.
Burton, J. T., of Toxside	1,240	£625
Caledonian Railway Company	39	281
Do. (Railway) . .	411	37,661
Callander, H., of Prestonhall, Guardians of .	4,869	6,810
Do. (Mines) . .	...	55
Campbell, Sir G., of Succoth, Bart. . . .	233	1,649
Carmichael, Sir W. H. G., of Durie, Bart. .	732	4,624
Christie, B., of Baberton	320	1,192
Clerk, Sir G. D., of Penicuik, Bart. . .	12,696	8,919
Do. (Mines) . .	...	2,421
Cochrane, A., of Ashkirk	243	1,173
Cochrane, J., of Harburn	4,064	1,946
Cowan, A., & Sons, papermakers . .	18	3,387
Cowan, Charles, of Loganhouse . .	5,677	1,816
Craig, Sir W. Gibson, of Riccarton, Bart. .	1,882	6,037
Craigcrook Mortification, Trustees of .	334	1,259
Cranston, G. C. T., of Dewar . .	1,652	632
Crown, The—War Department . . .	67	1,614
Cunyngham, Sir R. K. A. Dick, of Prestonfield, Bart.	228	1,759
Dalhousie, Earl of, Dalhousie Castle . .	1,419	3,002
Do. (Mines) . .	...	450
Davidson, Thomas, of Muirhouse . .	412	1,216
Dewar, Colonel, of Vogrie . . .	1,936	2,898
Dickson, J., of Corstorphine, Trustees of .	842	2,539
Dodds, John, Prestonkirk	1,200	309
Dougal, Mrs, of Ratho . . .	656	1,731
Douglas, J. D. S., of Baads . . .	3,106	1,388
Do. (Mines) . .	...	366
Dundas, Robert, of Arniston . .	10,184	9,549
Do. (Mines) . .	...	4,254
Durham, Mrs, of Polton . . .	666	1,738
Do. (Mines) . .	...	300
Edinburgh City Parochial Board . .	270	2,032
Edinburgh and District Water Trustees .	243	19,828
Elphinstone, Lord, Carberry Tower . .	769	2,580
Do. (Mines) . .	...	1,210
Fairholm, G. K. E., of Old Melrose . .	6,200	2,020
Fettes, Sir W., of Comely Bank, Bart., Trustees of	331	2,959
Forrest, Sir J., of Comiston, Bart. . .	500	1,290
Foulis, Sir J. L., of Colinton, Bart. . .	2,804	2,163
Frier, R. S., of Meikle Catpair . . .	1,025	800
Gartshore, J. M., of Ravelston . .	294	1,388
Gibsone, Lieut.-Gen. J. C. H., of Pentland .	1,474	2,338
Gilmour, W. L., of Craigmillar . .	1,690	7,960
Gordon, Capt. John, of Cluny . .	701	2,116
Graham, John, of Muldron, West Calder .	2,169	264
Do. (Mines) . .	...	363
Hare, S. B., of Calderhall . . .	2,373	3,107
Do. (Mines) . .	...	1,074
Hope, Sir A., of Craighall and Pinkie, Bart. .	961	3,436

Name of Landowner.	Estimated Acreage of Property.	Gross Annual Value.
Hunter, James, of Colzium	3,000	£750
Inglis, C. H. C., of Cramond . . .	637	2,520
Inglis, Right Hon. John, of Glencorse .	857	1,603
Inglis, Captain J., of Redhall . . .	712	1,937
Jeffrey, John & David	5	1,109
Johnston, George, of Lathrisk . . .	1,500	739
Lauder, Sir T. N. Dick, of Fountainhall, Bart. .	714	1,268
Lockhart, Sir S. M., of Lee, Bart. . .	700	1,327
Lothian, Marquess of, Newbattle Abbey .	4,547	11,918
Do. (Mines) . .	...	6,296
M'Dougal, T., Trustees of, Penicuik . .	62	1,472
M'Kelvie, James, Edinburgh . . .	7	1,240
Macfie, D. J., of Borthwickhall . . .	2,036	1,188
Macfie, R. A., of Dreghorn . . .	968	2,136
Maitland, Sir A. C. R. Gibson, of Cliftonhall, Bart.	4,505	14,246
Maitland, G. F., of Hermand . . .	567	1,426
Do. (Mines) . .	...	876
Melville, Viscount, Melville Castle . .	1,158	3,618
Miller, S. C., of Craigentinny . . .	652	5,739
Mitchell, Alexander, of Stow, Heirs of .	9,038	6,308
Morton, Earl of, Dalmahoy . . .	8,944	9,041
North British Railway Company . .	36	397
Do. (Railway and Canal)	873	35,634
Oakbank Oil Company, Mid-Calder . .	1	1,287
Do. (Mines) . .	...	300
Primrose, Rachel, of Burnbrae, Trustees of .	844	1,027
Raeburn, J. P., of Charlesfield, Heirs of .	497	1,338
Do. (Mines) . .	...	238
Ramsay, R. B. Wardlaw, of Whitehill .	2,963	3,822
Do. (Mines) . .	...	1,312
Ritchie, W., of Middleton . . .	2,652	3,137
Rosebery, Earl of, Dalmeny Park . .	15,568	8,973
Do. (Mines) . .	...	200
Scott, Major, of Malleny	3,250	3,964
Sivwright, Cathrine, of Southhouse, Heirs of .	497	1,724
Stair, Earl of, Oxenford Castle . . .	4,118	£3,165
Stair, Trustees of John, eighth Earl of .	8,384	4,988
Do. (Mines) . .	...	270
Stair, Earl of, and Trustees of John, eighth Earl of Stair	1,325	2,359
Stair family	13,827	£10,782
Stuart, J. & W., Musselburgh . . .	4	1,272
St Cuthberts, Parochial Board of, Edinburgh .	10	1,000
Thomson, G., of Burnhouse . . .	1,418	1,130
Tod, William, Lasswade . . .	13	1,227
Torphichen, Lord, Calder House . .	1,880	3,294
Do. (Mines) . .	...	500

Name of Landowner.	Estimated Acreage of Property.	Gross Annual Value.
Trotter, R., of Mortonhall	2,490	£6,759
Trotter, Colonel, of Bush	1,919	2,498
Do. (Mines) . .	...	500
Vere, J. C. Hope, of Craigiehall . . .	716	2,379
Walker, W. S., of Bowland	2,150	1,224
Watson, George, of Norton	470	1,381
Watson's Hospital, Governors of George . .	1,189	2,095
Wauchope, A., of Niddry	670	2,594
Do. (Mines) . .	...	300
Wauchope, Sir J. Don, of Newton Don and Edmonstone, Bart.	1,350	6,043
Do. (Mines) . .	...	267
Welwood, A. A. M., of Meadowbank . .	1,583	1,777
Wemyss, Earl of, Gosford House . . .	1,504	5,370
Do. (Mines) . .	...	200
White, W. L., of Kellerstain . . .	357	1,352
Wilkie, A., of Ormiston	1,579	2,289
Young, James, of Limefield	1,494	1,892
Do. (Mines) . .	...	2,266
Young's Paraffin Oil Company, West Calder .	331	6,580
Do. (Mines) . .	...	1,820

I. GEOLOGY.

THE leading geological features of Mid-Lothian are—1st, The great expanse of Calciferous Sandstones covering the west of the county; 2d, The large Coal basin in the east; 3d, The varied igneous rocks forming various hills; 4th, The absence of any but ancient rocks; and 5th, The deep coating of glacial clay which covers a large portion of the county. These features lead to many details of interest to the geological student; and whilst they will be treated in the most popular, concise, and elementary manner here, the man of science will find the rocks of the county skilfully and elaborately explained in the "Memoirs of the Geological Survey," particularly No. 32 ("The Geology of the Neighbourhood of Edinburgh," 1861); the late Charles Maclaren's classical treatise on the "Geology of Fife and the Lothians" (2d edition, 1866); and the "Transactions of the Edinburgh Geological Society" (3 volumes, 1868–77); whilst a host of writers of the greatest eminence, and belonging to various nationalities, have contributed to various learned societies at home and abroad papers on the rocks of the Edinburgh district, which, for various reasons, are celebrated all the world over.

Beginning at the lowest and ascending to the most modern rocks, we shall notice—

1st, *The Silurian Rocks.*—These very ancient rocks are found in patches on the Pentland Hills; for example, near the North Esk Reservoir. Two series, an upper and lower Silurian, have been well recognised from the distinctive fossils each contains. Proceeding to the Moorfoot Hills, we find them to be chiefly composed of rocks of lower Silurian age. Here, at the Moorfoots, we are not far from the centre of a great band of lower Silurian rocks, which stretches from St Abb's Head, in Berwickshire, across the south of Scotland to Portpatrick. The scenery of the Scottish Lowlands is mainly among Silurian rocks, which give rise to long undulating grassy hills, without striking or very picturesque features. In the vicinity of Edinburgh, however, the tameness of such scenery is quite redeemed by the charmingly rugged and picturesque trap-hills which abound there, and which shall afterwards be referred to.

2d, *Old Red Sandstone Rocks.*—These overlie the Silurian, and are consequently more modern, though also of vast antiquity. Indeed, with the exception of the surface-covering of glacial clay, all the rocks of the Edinburgh district belong to the most ancient geological period (the Palæozoic). We must go to England for abundant instances of modern rocks. The Old Red Sandstone rocks have likewise been divided into an upper and lower group. Red shales and conglomerates, and red and green sandstones and conglomerates, distinguish the Mid-Lothian rocks of the Devonian, or Old Red Sandstone age, from the Silurian shales and grits of the county. It has been suggested by Professor Geikie ("Geological Survey Memoir," No. 32) that, during the Old Red period, the Silurian shales and grits "rose above the level of the water as a group of low islands, round and over which sand and shingle accumulated to form the Old Red sandstone and conglomerate; and that the present Silurian patches (in the Pentlands) are the tops of some of the more prominent islands from which these deposits have been subsequently bared away." Old Red Sandstone rocks form the main body of the Pentland Hills, the felstones accompanying them belonging to the same geological period, and representing the eruptions of that now distant era.

3d, *Carboniferous Rocks.*—These are famous for their supplies of building-stone, coal, limestone, and shale-oil (paraffin). The value of mines of coal and shale to proprietors of land in Mid-Lothian is fully attested by the rentals of "mines" contained in the statistics of landed estates given in the preceding "General Observations." The value of the quarries of building-stone in Mid-Lothian is well shown by the stone of which the greater part of the New Town of Edinburgh is built, particularly that of "Craigleith Quarry," which yields a building-stone unequalled for beauty or durability. Owing, however, to its hardness, it is expensive to work, and is not now employed for building houses. Had Edinburgh been, like most English towns, dependent upon brick instead of stone, it would never have presented the splendid architectural appearance of to-day.

Like other geological formations, the Carboniferous has been divided into various successive groups of rocks. Lowest of all are found those Calciferous Sandstones which have yielded, with their associated rocks, such profitable commercial returns, besides increasing the rentals of the landowners. The *building-stone* furnished from this geological sub-formation has been already referred to. *Lime* of a very valuable kind is obtained from it at Burdiehouse on the east, and in the vicinity of East Calder and Newpark on the west of the Pentlands. Mr Torrance works the limestone at Camps Quarry, East Calder; and Messrs W. Baird and Co. work it at Westfield, at a depth of 32 fathoms, and Mr James Steel works it at Murieston, both near Newpark Railway Station, opencast. This limestone, both in the east and west beds, is of the best quality for iron smelting, &c. The now very important *shale* or *paraffin* oil industry at Oakbank, near Mid-Calder, and Addiewell (Young's Oil Co.), near West Calder, derives its supply of shale from beds of this age, the history of which, evidenced by fossils, seems to have been as follows:*—The opening chapters of the epoch reveal the district under the deep waters of the Carboniferous sea wherein were deposited those vast masses of sandstone out of which our houses and cities have been built. Igneous agencies, submarine and subærial, were soon at work occasioning those masses of dark traprock and basalt which we see around us, and whose rugged features lend picturesqueness to the landscape. Where the basins of limestone now exist at Burdiehouse and East Calder, great lagoons once stretched. Competent palæontologists have stated that these lagoons were of an estuarine character; that is, there must have been a junction here of a body of fresh water with the sea. The term "fresh-water limestone," which has been applied to that of Burdiehouse, is therefore erroneous. Slowly the deep estuaries or lagoons gave place to marshes and muddy pools, and the shale we so

---

* Address by the author to the Western District of Mid-Lothian Agricultural Association at Mid-Calder.—*North British Agriculturist*, 2d May 1877.

value to-day was the mud of that distant era. The mud was saturated with oil, chiefly from plant remains, occasionally from animal remains; and in the paraffin-lamps which illumine our houses, we burn an oil derived from the life of a remote past, a life which lay dormant and useless in the bowels of the earth until released from its shaly prisonhouse by the science of Lyon Playfair and the skill of James Young.

Above the Calciferous Sandstone group comes that of the Carboniferous Limestone, worked at Gilmerton, &c., consisting of marine (not estuarine) limestone and many igneous rocks, besides sandstones and shales. This group is developed chiefly in the eastern division of the county. It represents a period of great organic wealth, many of the limestones of this age teeming with fossil life. The "Millstone Grit," consisting of coarse red and white sandstones and conglomerate, follows.

At last the sea left large portions of the county dry, and soil formed, now represented by *fireclay*. Vegetation of the most luxuriant kind speedily followed, which, upon being afterwards entombed and chemically disintegrated, produced the *coal* for which the eastern district of the county has long been celebrated. This coal is itself divided into household and gas coal, in the latter of which especially the district excels. Coal, ironstone, and fireclay form the three great mineral products of this geological epoch.

We have now a vast hiatus in the geological record. Whilst in England seven formations succeed the Carboniferous, and mainly compose the land, in Mid-Lothian, and nearly all Scotland, there is nothing between the Carboniferous—one of the most ancient formations—and the Post-Tertiary, the most recent of all. Above the Carboniferous rocks we very frequently find in Mid-Lothian a more or less deep covering of glacial or boulder clay or till—a clay full of polished, striated, and travelled pebbles. How the till was formed is still a matter of dispute. There is little doubt, however, that the polishing and striation of its pebbles was due to the action of ice, which in one form or another once reigned here supreme, as the beautiful *roches moutonnées* of Corstorphine Hill testify. (See paper by author, "Edinburgh Geological Society's Transactions," vol. iii. part i.).

With regard to the *igneous* rocks of the county, which belong to various geological ages, Professor Judd's paper on the structure and age of Arthur's Seat deserves special notice. ("Quarterly Journal of the Geological Society of London" for May 1875). He says :—" At about the middle of the Calciferous Sandstone period (lower Carboniferous), a great fissure opened in the district, and began to emit at several points along its course discharges of vapour, carrying with them fragments of rock, cinders, and ashes, alternating with the outflow of currents of lava." This was the first appearance of the now famous hill of Arthur's Seat. "Eventually, by gradual elevation, this submarine volcano became a subaerial one, as is shown by the characters of the agglomerates composing the higher and later portions of its mass."

As to the other eminences throughout the county—Corstorphine Hill, Craiglockhart, the Dalmahoy Crags, &c.—they are composed of trap or greenstone, an ancient form of lava which burst out during Carboniferous times, and is found in various parts of Mid-Lothian. The Castle Rock is composed of a variety of basalt known as clinkstone, which pierced strata of sandstone, scorching and twisting them in its passage upwards, as may be seen from Johnston Terrace. The Calton Hill consists of a series of beds of trap-ash (with volcanic bombs), felstone, sandstone, and shale.

One of the most remarkable geological phenomena of the Edinburgh district (although Stirling Castle Rock and North Berwick Law are also instances of it) is what is termed *crag* and *tail, i.e.*, a crag or rocky hill with a tail of stratified rocks. The Castle Rock is an excellent example. It would appear as if some powerful agent had swept the country from west to east. The question is discussed in a paper by the author, "On the Rocks of Corstorphine Hill," published in 1877 in part i. of vol. iii. of the "Edinburgh Geological Society's Transactions."

On the whole, the county of Edinburgh presents to the geological student some of the finest examples of geological phenomena, whether as connected with igneous or stratified rocks, with physical or palæontological geology. It is a classical field to the geologist. Here Hutton founded a school which has numbered, and still numbers, many of the most eminent geologists in the world; and although some of his followers may have carried theories of " denudation," " glacial action," and " uniformity," farther than Hutton probably would, the " Edinburgh School of Geology," as it has been called, although its members are found in every part of the globe, still contains the most energetic, popular, and eloquent exponents of the science of Geology.

II. AGRICULTURE.

THE accompanying Agricultural map of the county of Edinburgh is, it is believed, the first attempt yet made on this side of the English Channel to convey to the eye, by means of colours on a map, the various agricultural rentals of a large tract of country. As has been explained, the idea is taken from Professor Delesse's " Carte agricole de la France, donnant la richesse agricole," published in 1874. There are two kinds of agricultural maps possible :—1. A map showing the various soils classified according to their various mineralogical and chemical constituents; or, 2. A map showing the various agricultural rentals. The latter kind of map was adopted by Professor Delesse, and has been followed here. The former kind of map should be prepared by the officers of H.M. Geological Survey as soon as they have mapped the various rocks which compose the framework on which the various soils rest. In the construction of the Agricultural map appended to this treatise, the author has had the invaluable assistance of Mr James Melvin, Bonnington, late President of the Scottish Chamber of Agriculture. The author begs an indulgent criticism of this map, and requests that corrections will be communicated to him for a future edition.

A word as to the connection between rocks and the soils lying above them. There can be no doubt that surface-soil is largely the result of the action of water and the weather upon the subjacent rocks. When vegetation, however, once begins, the death of successive generations of plants, and the action of animals increases the surface-soil; whilst the presence of the vegetation protects the rocks from much weather action, although not from disintegration by water in its various forms. Looking to the connection merely between rocks and the soil, there is undoubted evidence that the disintegration of certain rocks produces bad, whilst that of others produces good soils. Thus soil formed from, and resting upon immediately subjacent shale, will be a poor soil; soil formed from subjacent glacial clay will be a wet, cold soil; soil formed of disintegrated trap-rock will be a rich and fertile soil. A Geological map, however, gives no indication of the surface-soils of the country,—only of the rocks beneath. Thus, if the Geological Survey map of Mid-Lothian be compared with the Agricultural map of the county appended to this treatise, no connection between them will be apparent; for the Survey map gives only the solid rocks which form the upper crust of this part of the earth, and does not give the universal and often deep coverings of clay, sand, gravel, and other loose materials upon which the character and value of the surface-soil generally depend.* Such "universal coverings" form the "sub-soil" which augments and influences the "surface-soil." If, however, we had a map of the various soils of the county classified according to their mineralogical and chemical constituents, we should doubtless discover—keeping in view altitudes, temperatures, rainfalls, and the distances of localities from railways—a close connection between such a map and that appended to this volume. At the request of the French Government, the Highland and Agricultural Society of Scotland has prepared for the forthcoming International Agricultural Congress, to be held in connection with the Paris Exhibition of 1878, a series of short articles upon various subjects relating to Scotland of interest to agriculturists. The author was commissioned by the Society to give a sketch of the geology of Scotland, and he therein takes the opportunity of pointing to the immense difference between an agricultural map, such as accompanies this treatise, and purely geological maps, such as have been prepared by H.M. Geological Survey. If the surface-soils of the United Kingdom were systematically classified and mapped, a work of no ordinary economic importance would be achieved.

* "The defect," says a writer in the *Journal of Agriculture* for January 1857, "of all existing geological maps, for Agricultural purposes, is the absence of correct information regarding the Northern Drift and superficial accumulations, which often exercise a much greater influence on the character of the soil than the fixed rock formations."

Agriculturally, the county of Edinburgh takes a high position. Not only have her farmers brought the science of agriculture to a standard of perfection rarely attained, but the agricultural produce and value of land in the better portions of the county are exceptionally good. Of the county's total area of 234,926 acres, 133,465 were in 1876 under crops, barefallow, and grass. In 1877 this had increased to 134,138 acres. These figures are obtained from the Government Agricultural Returns published in 1877. The following will exhibit to what particular crops the cultivated acreage in 1876 and 1877 was devoted, the returns having been made to Government on 25th June 1876 and 4th June 1877:—

MID-LOTHIAN.—Total area, 234,926 acres.

	1876.	1877.
Total acreage under Crops, Barefallow, and Grass,	133,465	134,138
Corn Crops :—		
Wheat,	4,456	4,966
Barley or Bere,	11,982	11,811
Oats,	21,311	22,221
Rye,	31	35
Beans,	347	339
Peas,	62	45
Total of Corn Crops,	38,189	39,417
Green Crops:—		
Potatoes,	6,930	7,063
Turnips and Swedes,	13,343	12,987
Mangold,	19	14
Carrots,	43	18
Cabbages, Kohl-rabi, and Rape,	597	677
Vetches and other Green Crops, except Clover or Grass,	811	878
Total of Green Crops,	21,743	21,637
Clover, Sanfoin, and Grasses under rotation :—		
For hay,	11,874	11,275
Not for hay,	19,995	19,841
Total of Clover, &c.	31,869	31,116

	1876.	1877.
Permanent Pasture or Grass not broken up in rotation (exclusive of heath or mountain land):—		
For hay,	1,234	1,492
Not for hay,	40,283	40,198
Total of Permanent Pasture, &c.	41,517	41,690
Barefallow or uncropped arable land,	147	278

The following are the Government returns of the numbers of live stock in Mid-Lothian, as returned upon 25th June 1876 and 4th June 1877 :—

	1876.	1877.
Horses (including ponies), as returned by occupiers of land :—		
Used solely for purpose of agriculture, &c.,	3,399	3,401
Unbroken Horses and Mares kept solely for breeding,	806	660
Total of Horses,	4,205	4,061
Cattle :—		
Cows and Heifers in milk or in calf,	9,514	10,187
Other cattle :—		
2 years of age and above,	4,281	4,010
Under 2 years of age,	4,866	4,132
Total of Cattle,	18,661	18,329
Sheep :—		
1 year old and above,	103,222	101,411
Under 1 year old,	65,343	62,767
Total of Sheep,	168,565	164,178
Pigs—	5,158	5,765

In Mid-Lothian, in 1877, there were 94 acres of orchards, 865 of market gardens, 382 of nursery gardens, and 10,320 of woods (coppices or plantations).

Very ample agricultural statistics (procured under probably more favourable circumstances than those enjoyed by Government enumerators) will be found in the Reports of the late Mr Hall Maxwell, Secretary to the Highland and Agricultural Society, for the years 1854, 1855, 1856, and 1857. Since 1857, the Agricultural Statistics of Scotland have been published solely by Government. The author is indebted to Mr F. N. Menzies, Secretary of the Highland and Agricultural Society of Scotland, for the statistics referred to in this article. Let us compare Mr Hall Maxwell's returns for the county for 1855 and 1856 with the Government returns for 1876 and 1877:—

Crops, &c.	1855. Acres.	1856. Acres.	1876. Acres.	1877. Acres.
Under a Rotation of Crops, .	102,604	104,077	...	...
„ Crops, Barefallow, and Grass, . .	...	...	133,465	134,138
„ Wheat, ...	9,113	11,628	4,456	4,966
„ Barley or Bere, . .	10,460	10,222	11,982	11,811
„ Oats, ...	23,488	23,181	21,311	22,221
„ Rye, ...	$\frac{1}{4}$	8$\frac{1}{4}$	31	35
„ Beans, . . .	1,426	1,576	347	339
„ Peas, ...	144	181	62	45
„ Potatoes, . . .	6,749	6,668	6,930	7,063
„ Turnips, . . .	13,862	14,517	13,343	12,987
„ Mangold, . . .	96	124	19	14
„ Carrots, . . .	38	91	43	18
„ Cabbages and Rape, .	75	109	597	677
„ Vetches, . . .	1,135	1,263	811	878
„ Grass and Hay under Rotation, . .	35,304	33,703	31,869	31,116
„ Barefallow, . . .	441	497	147	278
„ Permanent Pasture or Grass not under Rotation, . . .	...	...	41,517	41,690

Stock.	1855.	1856.	1876.	1877.
Total number of Horses, . .	5,002	4,994	4,205	4,061
„ Cattle, . .	16,902	16,078	18,661	18,329
„ Sheep, . .	147,543	143,470	168,565	164,178
„ Pigs, . .	5,225	5,266	5,158	5,765
Total of Stock, .	174,672	169,808	196,589	192,333

	1855–6.	1856–7.	1876–7.	1877–8.
Valued Rent of County (exclusive of Railways),	£402,064	£409,607	£558,194	£578,583

	1851.	1861.	1871.
Population of County according to Census, . . .	259,435	273,997	328,379

With regard to the wages of farm servants in Mid-Lothian, Mr Thomas Farrall states in his Prize Essay ("Transactions of the Highland and Agricultural Society" for 1877) that at present the wages of ploughmen range from £51 to £53 per annum;* ordinary labourers receive from 16s. to 20s. per week, and women workers 9s. per week. In 1795, hired men received £6, 10s., and women £3, 10s. per annum; and day labourers were paid 6d. a day in winter and 8d in summer, with victuals. "Wages have gone up fully 50 per cent. in twenty-five years, 40 per cent. in ten years, and £10 per annum in five years." The *working expenses* of large farms near Edinburgh are estimated at from 38s. to 45s. per acre, a rise of 50 per cent. in 25 years.

As to the *size of farms* in Mid-Lothian, Mr Farrall gives the following table as showing the various tenancies into which the county is divided :—

50 Acres and under.	50 to 100 Acres.	100 to 300 Acres.	300 to 500 Acres.	Above 500 Acres.	TOTAL.
477	116	294	75	50	1012

Ordinary farms in Mid-Lothian range from 100 to 400 acres, some occupations in the hill districts being still larger. Unless under exceptional circumstances, farms of less than 100 acres are not considered remunerative. The leases of farms properly so-called, are all for nineteen years; this term being likewise considered essential for good and successful farming.

As to the *rental* of land, the accompanying agricultural map of the county shows how much it varies, according to the nature of the soil and the favourable position of the farm—farms becoming more and more highly rented as the metropolis is approached. In the immediate vicinity of the city of Edin-

* Estimated as follows—Say £37 in cash; free house and garden; 4 bolls of potatoes; 6½ bolls of oatmeal; and coals driven. Equal in all to about £1 per week.

burgh lands acquire an extraordinary and artificial value from the following causes—1. Highly-rented fields are leased by city butchers for grazing cattle. 2. By means of irrigation with the city sewage, lands, such as those of Craigentinny, have acquired a fabulous value.* 3. Vegetable market-gardeners and nurserymen occupy considerable portions of highly-rented ground near the city. 4. Competition with the builders of the suburbs of Edinburgh, a city which has been rapidly increasing of late years, has given an artificial value to the land immediately surrounding the city. At the same time, there is no doubt that the soil in the vicinity of the metropolis is very favourable to agriculture when scientifically pursued; and that, as compared with most other places in the county, and indeed in Scotland, the city of Edinburgh enjoys a drier climate, a most important consideration to the farmer.

The *Fiars prices* of the county of Edinburgh (that is, the average prices, yearly ascertained by the Sheriff, with the help of a jury, of the several kinds of grain grown in the county in the preceding crop) are stated in *Oliver & Boyd's Almanac* as follows :—

	1870	1871	1872	1873	1874	1875	1876	Average of seven years.
Wheat, 1st, p. Qr.	51/	50/1	34/	55/2	40/3	40/	42/2	44/8
,, 2d, ,,	48/	47/	33/	51/	38/	37/6	39/6	42/
Barley, 1st, ,,	34/6	36/5	30/	42/1	39/11	35/4	32/9	35/10¾
,, 2d, ,,	30/6	34/	28/	40/	36/	30/6	29/	32/6⅝
,, 3d, ,,	28/	30/	26/6	38/	35/	...	....	...
Oats, 1st, ,,	26/	26/5	26/1	28/11	29/9	29/	27/7	27/8¼
,, 2d, ,,	23/	24/	25/10	27/	28/	27/	25/	25/8¾
Peas and Beans,	45/	43/	39/	44/	27/6	46/10	42/11	41/2¼
Oatmeal, p. Boll of 140 lbs.	20/	19/9½	21/	21/9	21/11¾	20/9	21/3	20/11 $\frac{2}{13}$

Comments on these prices as a test of the agriculture of Mid-Lothian are made in next article.†

* The article "Irrigation" in *Chambers' Encyclopædia* thus refers to them: "The most successful instance of sewage irrigation in Great Britain is to be found near Edinburgh, where an extensive tract of meadows lying between Portobello and Leith yield an annual rent of £20 to £40 an acre; the grass is cut from three to five times a year, and as much as ten tons an acre have been obtained at a cutting."

† The Fiars prices for 1877 (struck 8th March 1878) are given at p. 36.

III. METEOROLOGY.

In the more level seaboard districts of the county of Edinburgh the climate is mild and dry; and this, added to the favourable soil found there, has undoubtedly been the cause of its agricultural superiority. Among the hilly districts of the county, again, other conditions prevail. Not only is the soil poorer, but the climate is much more moist. These considerations, agricultural and meteorological, have led to the county being divided—for the purpose of striking the fiars prices by jury, as mentioned at the close of the preceding article—into three agricultural districts: Low, Middle, and High Mid-Lothian, the average prices for the three districts being returned as the fiars prices of the county. The low district runs along the sea-coast, the middle is between the low district and the hills, and the high district comprises hill farms. In this way the fiars prices of Mid-Lothian do not represent the prices of grain from the best land; and for this reason, too, the fiars prices of a county like that of Edinburgh are actually lower than those of some counties which have a greater proportion of superior to inferior land, but which possibly have not such fine land as some parts of Mid-Lothian. The county of Edinburgh consists of land of all degrees of excellence, situated at very varying elevations above the sea-level, and subject to very different temperatures and rainfalls. In this article the author is indebted to the statistics furnished by the Scottish Meteorological Society, which has numerous stations of observation throughout the county, all of which will be found on reference to the accompanying Agricultural Map. These statistics appeared in vol. iii., Nos. 28, 29 (1871), and 33, 34 (1872) of the Society's Journal, and were prepared by its meteorological secretary, Mr Alexander Buchan, M.A., who has also kindly favoured the author with some hitherto unpublished observations.

It is proposed to limit this article on the meteorology of the county of Edinburgh to a consideration simply of its various temperatures and rainfalls at various elevations, these three constituents, (1) Temperature, (2) Rainfall, and (3) Altitude, being the main meteorological topics of agricultural interest. For, to quote Mr Buchan, " a knowledge of the temperature required for the ripening of cereals is most important to the agriculturist. It has been proved by observations made by those who competed for the prizes offered by the Marquess of Tweeddale, President of the Scottish Meteorological Society, that for the ripening of wheat and barley, with the ordinary range of temperature in Scotland, there must be an average temperature of 56°. In England, with its shorter day in summer, a higher temperature is doubtless required, since on the Continent, with its still shorter day, 58° is required. If the temperature falls below this, there is a deficiency in the crop; but if it rises above it, the crop is so much the better, other conditions being favourable."

Turning first, then, to the temperature of the county, we shall find it pretty accurately from the observations made at five stations of the Scottish Meteorological Society:—

Stations.	Height of Station above Sea.	No. of Years ending with and including 1869) for which Observations taken.	Mean Temperature (in degrees Fahrenheit) for												Mean for Year.
			January.	February.	March.	April.	May.	June.	July.	August.	September.	October.	November.	December.	
	Feet.														
Edinburgh (Newington), .	270	8¾	36·3	39·1	40·0	45·1	50·6	55·6	58·3	57·7	54·0	47·7	40·9	39·8	47·1
Leith (Restalrig), .	80	13	38·9	39·9	40·8	45·0	49·2	55·5	58·2	57·8	54·6	47·1	41·3	40·5	47·4
Inveresk, . .	90	13	37·2	38·7	40·2	46·1	52·2	57·5	59·5	58·6	54·7	47·0	40·0	39·1	47·6
Dalkeith, . .	190	4	37·4	39·1	40·6	46·1	51·6	56·7	59·4	58·2	54·5	47·1	39·8	39·1	47·5
North Esk Reservoir (Pentlands),	1150	8	33·7	34·8	35·9	41·5	46·7	51·9	53·8	53·1	50·8	43·4	36·8	36·6	43·3

Dividing the year into four quarters, and comparing the temperature of Edinburgh with that of London (details given p. 29), we shall find the mean temperature of the seasons there to be as follows :—

	Edinburgh.	London.
SPRING (March, April, and May), . . .	45·2	48·8
SUMMER (June, July, and August), . . .	57·2	62·9
AUTUMN (September, October, and November), .	47·5	51·1
WINTER (December, January, and February), .	38·4	40·0
MEAN ANNUAL TEMPERATURE, . .	47·1	50·7

Comparing the temperature of Mid-Lothian with that of other places at home and abroad, we may adopt the mean annual temperature of the city of Edinburgh shown above, 47·1°, as fairly representative of that of the county. In this view, let us first compare with it some places in Scotland :—

Stations.	Feet above Sea.	No. of Years ending 1869.	Mean Temperature (in degrees Fahrenheit) for												Mean for Year.
			January.	February.	March.	April.	May.	June.	July.	August.	September.	October.	November.	December.	
Edinburgh, .	270	8¾	36·3	39·1	40·0	45·1	50·6	55·6	58·3	57·7	54·0	47·7	40·9	39·8	47·1
Glasgow, . .	180	13	37·7	38·7	39·9	45·3	50·6	56·3	58·0	57·3	53·6	47·2	40·3	39·9	47·1
Greenock, . .	64	13	39·1	39·6	40·6	46·0	51·2	57·3	59·0	58·3	54·4	48·2	41·8	41·2	48·1
Aberdeen, . .	110	13	37·3	38·3	39·7	44·6	49·2	55·7	57·6	57·2	53·7	46·9	40·7	39·1	46·7
Elgin, . . .	50	13	37·1	38·6	39·4	45·1	50·1	56·1	57·7	57·3	53·8	46·8	40·3	40·2	46·9
Perth, . . .	66	9	36·9	39·2	40·5	45·7	50·8	55·7	59·3	57·8	53·8	47·3	40·3	39·8	47·2
Girvan (Ayr), .	27	7½	39·7	40·7	41·0	46·7	51·7	56·3	58·2	57·9	55·0	49·4	43·0	41·3	48·4
Dumfries, . .	180	5	37·6	39·3	40·4	45·5	51·4	56·2	58·7	58·3	54·5	48·3	40·6	40·1	47·6
Stirling, . .	233	3½	37·9	38·2	39·5	45·2	49·7	56·0	57·0	56·9	53·9	46·7	40·9	40·5	46·9
Oban, . . .	48	5½	39·0	39·7	40·4	45·7	50·4	54·7	56·9	56·6	53·4	47·1	42·1	41·1	47·3
Eyemouth, . .	16	3½	37·8	38·9	39·9	44·7	49·8	55·7	57·7	57·8	54·6	48·3	42·1	40·6	47·7
East Linton, .	90	13	38·1	39·4	40·7	45·1	50·1	56·0	58·2	57·9	54·8	48·1	41·1	39·4	47·4

Leaving Scotland, and comparing the temperature of the city of Edinburgh with that of towns in the United Kingdom and abroad, we obtain the following results:—

Stations.	Feet above Sea.	No. of Years ending 1869.	Mean Temperature (in degrees Fahrenheit) for												Mean for Year.
			January.	February.	March.	April.	May.	June.	July.	August.	September.	October.	November.	December.	
Edinburgh (Newington),	270	8¾	36·3	39·1	40·0	45·1	50·6	55·6	58·3	57·7	54·0	47·7	40·9	39·8	47·1
London (Camden Town),	123	12½	38·5	40·9	42·4	49·0	55·1	61·3	64·6	62·9	59·0	51·8	42·5	40·6	50·7
Dublin,	159	13	40·6	41·9	42·4	47·6	52·0	57·2	59·7	58·8	55·5	50·8	43·3	42·3	49·3
Liverpool,	37	13	40·5	42·1	42·6	48·6	54·0	59·3	62·4	61·6	57·9	52·2	44·2	42·6	51·0
York,	50	13	36·6	38·9	40·4	46·6	51·9	58·2	60·2	59·8	55·5	49·4	41·7	39·0	48·2
Oxford (Radcliffe Observatory),	210	13	38·4	40·6	41·9	48·2	53·6	59·5	62·1	61·1	57·3	50·9	42·2	40·6	49·7
Clifton	228	12	39·2	41·0	42·2	48·9	53·9	59·6	62·4	61·5	57·7	51·4	42·9	41·6	50·2
Bournemouth,	100	8	40·5	42·3	43·2	49·0	53·7	58·9	62·2	61·4	58·5	53·1	45·0	43·2	51·2
Torquay,	150	7	42·4	42·8	44·5	48·3	53·7	59·6	61·2	60·0	58·4	52·9	45·5	43·5	51·1
Helston (Cornwall),	106	13	45·1	46·0	46·5	51·4	55·7	60·5	63·4	62·8	60·0	55·0	48·4	46·7	53·5
Ventnor (Isle of Wight),	153	10	41·6	43·1	44·3	49·8	54·7	59·2	62·6	62·0	59·8	55·0	46·4	44·2	51·9
Jersey (Millbrook),	50	6	42·8	43·8	45·0	51·0	54·7	59·3	62·4	62·3	59·7	55·0	47·8	45·2	52·4
Paris,	216	13	37·4	40·1	43·4	51·9	57·9	63·0	66·3	65·2	60·9	52·4	42·4	38·8	51·6
Brussels,	186	13	36·2	40·0	42·6	50·8	57·2	63·5	65·7	64·8	60·6	52·5	42·4	38·7	51·2
Copenhagen,	12	13	31·6	31·1	34·4	42·2	51·1	59·9	62·6	61·5	55·9	47·5	38·1	34·3	45·9
Iceland (Reykjavik),	10	4	30·4	27·4	27·8	35·3	43·5	46·7	50·9	50·5	45·3	38·5	33·7	31·0	38·2

The high elevation of the meteorological station at Edinburgh (Newington) might be supposed to lead to the temperature for that city being returned as colder than that of most English towns; but we must remember that even at the Leith station (Restalrig), which is only 80 ft. above the sea, the mean annual temperature was only 47·4°. There can therefore be no doubt that Edinburgh and Mid-Lothian generally are subject to a colder temperature than most English towns. At the same time, this is not peculiar to the county of Edinburgh; for it will be recollected that the mean annual temperature of Glasgow (only 180 ft. above the sea) was precisely the same as that of Edinburgh (270 ft.). The truth is, that over Scotland generally a colder climate prevails than over England generally. Out of the 76 stations, situated all over Scotland, of the Scottish Meteorological Society, not one, in the statistics referred to, registered during June, July, August, or September, a mean monthly temperature of 60°; whereas out of the 67 English stations, 17 during June, 58 during July, 55 during August, and two (Gloucester 68·4, and Helston 60·0) during September, had a mean temperature of 60° and upwards. Keeping in view Lord Tweeddale's observation, as to 56° being required in Scotland, and 58° in England as an average temperature for the ripening of wheat and barley, such meteorological returns as have been given acquire an important agricultural interest. The ripening of fruit is also similarly explained.

With regard to the *occurrence of frost* in the Edinburgh district during winter and spring, Mr James M'Nab states (" Transactions Edinburgh Botanical Society, 1877 ") that at the Royal Botanic Garden, Edinburgh (68 ft. above sea-level), of which he is Curator, the thermometer, in the morning, registered freezing (32° F.) or above, during—

							Greatest cold.
October 1876, not once,	.	.	.	.	.	33°
November „ 13 times,	.	.	.	.	.	20°
December „ 12 „	.	.	.	.	.	24°
January 1877, 16 „	.	.	.	.	.	18°
February „ 7 „	.	.	.	.	.	20°
March „ 20 „	.	.	.	.	.	17°
April „ 10 „	.	.	.	.	.	23°
May „ 6 „	.	.	.	.	.	21°
June „ not once,	.	.	.	.	.	37°

The weather during the spring and summer of 1877 was exceptionally cold and backward. " On July 12, it was impossible to procure in the fields near Edinburgh specimens of wheat, oats, and barley in a state fit for examination at the University Botanical class, a circumstance of rare occurrence at this date."

Let us next turn our attention to another topic of interest to farmers and the public generally, viz., the amount of Rain which falls in various parts of the county, and elsewhere. This rainfall is calculated by inches. One inch, for example, signifies that if the rain that fell remained standing upon the ground, it would cover it one inch deep.* In the county of Edinburgh, the Scottish Meteorological Society has, or had, no fewer than thirteen stations for registering rainfalls. We thus obtain the following returns representative of the rainfall of Mid-Lothian :—

* " An inch of rain represents about 100 tons of water to the acre." (Professor Huxley. *Physiography.* 1878).

MIDLOTHIAN RAINFALL.

STATIONS.	Height of Station above Sea.	No. of Years.	Years specified.	AVERAGE RAINFALL (IN INCHES) FOR												
				Jan.	Feb.	Mar.	Apr.	May.	Jun.	Jul.	Aug.	Sep.	Oct.	Nov.	Dec.	Year.
	Feet.															
Edinburgh,	Various.	50	1822-71	1·81	1·61	1·55	1·36	1·68	2·38	2·68	2·70	2·33	2·50	2·07	1·94	24·61
Do.,	230	22	1850-71	2·50	1·89	1·64	1·54	1·95	2·49	2·40	2·66	2·50	2·43	2·02	2·38	26·40
Inveresk,	90	35	1837-71	2·17	1·77	1·58	1·51	1·84	2·61	2·51	2·69	2·43	2·62	2·27	2·07	26·27
Meadowfield (Corstorphine),†	155	19	1853-71	2·24	1·60	1·44	1·54	1·66	2·28	2·30	2·35	2·23	2·32	1·81	1·97	23·75
Bonnington (Ratho),	405	22	1850-71	2·66	2·04	1·62	1·43	1·69	2·41	2·60	2·84	2·27	2·72	2·12	2·56	27·01
Fernielaw (Colinton),	500	22	1850-71	3·89	2·84	2·24	2·07	2·14	2·92	2·85	3·14	2·83	3·09	3·12	3·27	34·40
Swanston,	555	23½	June 1848-71	3·70	2·65	2·05	2·00	2·22	2·99	2·73	3·15	2·97	3·58	2·89	2·97	33·90
Clubbiedean,	780	12	1860-71	4·25	2·32	2·82	2·23	2·34	2·55	3·36	3·01	3·18	4·11	3·12	3·16	38·50
Glencorse,	787	41	1831-71	3·49	2·93	2·68	2·03	2·17	3·06	3·28	3·32	3·24	3·85	3·10	3·23	36·38
Harlaw,	800	18	1854-71	3·69	2·64	2·35	2·30	2·35	2·87	2·98	3·08	3·31	3·93	2·98	3·21	35·69
Cobbinshaw,	800	13	1862-74	4·00	2·90	2·48	1·89	2·32	2·12	3·02	3·45	3·58	4·13	3·46	3·31	36·66
Colzium,	1000	13	1848-60	4·25	3·44	2·86	2·50	2·28	4·05	3·73	3·55	3·23	4·63	3·38	4·62	42·52
North Esk Reservoir,	1150	21	1851-71	4·54	3·43	2·44	2·36	2·17	2·96	3·01	3·33	3·32	4·19	2·75	3·99	38·49

(Pentland Hills.* — bracketed together: Swanston, Clubbiedean, Glencorse, Harlaw, Cobbinshaw, Colzium, North Esk Reservoir)

* These rain-gauges are kept by the officials of the Edinburgh and District Water Trust, to estimate the water supply of Edinburgh derived from reservoirs in the Pentland Hills. The latest returns for Glencorse are :—1875, 36·70; 1876, 45·80; 1877, 54·35 inches of rain, showing an annual increase during the past three years.

† This is the driest station in Scotland (*vide* p. 34 for the wettest).

Judging from the above returns, April was, with only three exceptions, the month during which least rain fell. As to the month when most rain fell, January, August, and October equally merit this unenviable distinction. Dividing the year into four quarters, and comparing the rainfall of Edinburgh with that of London (details given p. 35), we shall find the rainfall of the respective seasons to be there as follows :—

	Edinburgh.	London.
Spring (March, April, and May), . .	4·59	5·32
Summer (June, July, and August), . .	7·76	6·75
Autumn (September, October, and November), .	6·90	7·32
Winter (December, January, and February), .	5·36	5·27
Mean Annual Rainfall,	24·61*	24·66

Taking the rainfall of the counties of Edinburgh and Haddington together, statistics prove that the lower and most agriculturally valuable districts of these counties are the driest in all Scotland. This satisfactory result is obtained from the united rainfall returns of Bonnington, Meadowfield, Edinburgh (two stations), Inveresk, Tyneholm, Haddington, Smeaton, East Linton, and Thurston. The area embraced by those stations is at once the driest in Scotland, and the portion of North Britain where agriculture has been most successfully and scientifically pursued. Passing to other districts of Scotland, we find the rainfall often so great as to seriously interfere with agriculture. The following instructive Table by Mr. Buchan gives the various rainfalls of various Scotch districts :—

* This is the average for the years 1822-71, that of London being for the years 1815-69. Long periods like this should be compared with long periods only.

AVERAGE RAINFALL (IN INCHES) FOR EACH MONTH.

SCOTCH DISTRICTS.	Jan.	Feb.	Mar.	Apr.	May.	June.	July.	Aug.	Sept.	Oct.	Nov.	Dec.	Year.
EAST OF SCOTLAND.													
Round Moray Firth,	2·55	2·37	1·85	1·77	1·58	1·84	2·31	2·79	2·94	3·14	2·67	2·77	28·58
North-Eastern (Aberdeen, &c.),	2·95	2·46	2·43	2·00	1·73	1·76	2·18	2·96	2·84	3·50	3·12	3·25	31·18
Kincardine and Forfar,	3·01	2·20	2·02	1·97	1·87	2·27	2·58	2·71	2·76	3·39	2·61	3·06	30·45
East Perthshire,	3·97	2·89	2·26	1·91	2·00	2·46	2·98	3·16	3·02	3·43	2·62	3·63	34·33
Mid and East Lothian,	2·21	1·68	1·58	1·48	1·77	2·35	2·46	2·52	2·44	2·67	2·15	2·05	25·36
Teviot and Lower Tweeddale,	3·47	2·55	2·46	2·13	2·35	2·25	2·68	2·72	3·08	3·68	2·50	3·49	33·36
WEST OF SCOTLAND.													
North-Western (Tongue, &c.),	6·64	5·39	4·66	3·13	2·64	2·96	3·61	4·27	5·26	6·13	5·88	7·12	57·69
West Argyll,	6·58	5·62	4·19	3·58	3·13	3·35	3·92	5·10	5·16	6·00	5·07	7·26	59·11
Lower Clydesdale,	3·83	3·19	2·63	2·15	2·30	2·81	3·22	3·48	3·42	3·82	2·99	3·72	37·56
Renfrewshire,	6·01	4·97	3·69	2·86	2·76	3·13	3·45	4·24	4·22	5·31	4·09	5·94	50·67
W. Ayr and Wigtown,	4·98	4·01	3·37	2·80	2·62	3·42	3·71	4·18	3·76	4·90	4·25	4·54	46·54
Nithsdale,	4·99	3·68	2·79	2·26	2·39	2·42	2·55	3·54	3·51	4·50	3·17	4·68	40·48
Eskdale (Langholm, &c.),	6·31	4·54	3·86	2·89	3·05	3·45	3·60	4·91	4·79	5·60	4·21	5·70	52·91
NEAR HILLS.													
Lead and Moffat Hills,	7·50	5·44	5·20	3·77	4·03	3·95	4·10	6·06	5·96	6·00	5·07	7·96	65·64
Lammermoors,	2·94	2·24	2·10	2·00	1·97	2·44	2·31	2·50	3·01	3·56	2·68	2·90	30·65
Pentlands,	3·07	2·89	2·49	2·22	2·24	3·06	3·13	3·23	3·15	3·91	3·05	3·49	36·83
Ochils,	4·36	3·06	2·30	2·08	1·98	2·69	3·13	3·42	2·91	3·78	2·79	4·05	36·55
Perthshire, &c., Hills,	7·97	6·30	4·74	4·06	3·74	4·62	4·72	6·50	6·16	7·81	5·78	8.17	70·57

Let us now compare the rainfalls of some towns throughout Scotland :—

STATIONS.	Height of Station above Sea.	No. of Years.	Years specified.	AVERAGE RAINFALL (IN INCHES) FOR												
	Feet.			Jan.	Feb.	Mar.	Apr.	May.	Jun.	Jul.	Aug.	Sept.	Oct.	Nov.	Dec.	Year.
Edinburgh,	230	22	1850-71	2·50	1·89	1·64	1·54	1·95	2·49	2·40	2·66	2·50	2·43	2·02	2·38	26·40
Glasgow,	180	15¼	July 1856-71	4·77	4·10	3·39	2·54	2·43	2·93	3·27	3·97	3·68	4·21	3·13	4·48	42·90
Greenock,	700	36	1835-71	7·06	6·01	4·59	3·33	3·15	3·93	4·16	4·89	4·76	6·04	5·67	7·09	60·68
Portree (Isle of Skye),	60	12	1860-71	12·00	10·49	7·40	5·43	4·51	4·50	5·75	6·82	8·94	10·01	11·00	12·48	99·33
Aberdeen,	99	34	1838-71	2·57	1·90	1·92	1·82	1·61	1·99	2·27	2·68	2·39	3·11	2·80	2·79	27·85
Perth,	66	16	1856-71	4·62	3·31	2·62	2·22	2·17	2·87	3·51	3·31	3·22	3·55	2·85	3·73	37·98
Dumfries,	70	22	1850-71	4·18	2·76	2·25	2·03	2·06	2·45	2·51	3·58	2·81	3·84	2·84	3·77	35·08
Oban,	10	8¾	{1859-65 & 1870-71}	7·77	5·80	5·76	3·27	3·55	3·81	3·87	5·92	5·94	6·90	5·47	6·12	64·18
Upper Glencroe (near head of Loch Fyne, Cor-)	520	6½	June 1864-70	16·84	15·63	7·48	8·82	6·48	6·88	6·84	7·99	11·99	12·37	10·59	15·92	127·65
Meadowfield, storphine, near Edinburgh,	155	19	1853-71	2·24	1·60	1·44	1·54	1·66	2·28	2·30	2·35	2·23	2·32	1·81	1·97	23·75

The two last-quoted stations are respectively the wettest and driest in Scotland.

Finally, in order to compare the rainfalls of places in England, Ireland, and abroad, with the rainfall of Mid-Lothian, the following hitherto unpublished statistics, kindly supplied by Mr Buchan, are given:—

STATIONS.	YEARS SPECIFIED.	AVERAGE RAINFALL (IN INCHES) FOR												
		Jan.	Feb.	Mar.	Apr.	May.	June.	July.	Aug.	Sept.	Oct.	Nov.	Dec.	Year.
Edinburgh,	1822–71	1·81	1·61	1·55	1·36	1·68	2·39	2·68	2·70	2·33	2·50	2·07	1·94	24·61
London (Greenwich),	1815–69	1·80	1·57	1·55	1·65	2·12	1·90	2·56	2·29	2·36	2·73	2·23	1·90	24·66
Manchester,	1793–71	2·54	2·41	2·29	2·06	2·30	2·86	3·56	3·50	3·32	3·89	3·78	3·29	35·50
Gloucester,	1859–71	2·25	1·29	1·67	1·61	2·14	2·16	2·11	2·09	2·98	2·46	1·79	1·98	24·53
Oxford,	1828–71	1·92	1·49	1·47	1·69	1·87	2·39	2·48	2·38	2·66	2·65	2·10	1·69	24·79
Bodmin (Cornwall),	1850–71	5·74	3·03	3·42	2·82	2·87	2·92	3·04	3·32	3·81	5·43	4·41	4·90	45·71
Dublin,	1837–52 and 1855–69	2·26	1·80	1·85	2·12	2·22	2·19	2·50	2·73	2·30	2·85	2·51	2·16	27·69
Cork,	1857–71	4·21	2·51	2·87	2·44	2·27	1·98	2·03	2·70	3·18	3·57	3·36	4·41	35·53
Paris,	1805–70	1·47	1·42	1·21	1·25	1·44	1·93	1·90	2·02	1·79	1·94	1·79	1·72	19·80

In concluding this little treatise on the Geology, Agriculture, and Meteorology of the county of Edinburgh, the author would again thank the scientific friends who have enabled him to cite the most reliable statistics; and he trusts the public will regard indulgently an essay upon subjects of vast importance, although contained in the narrow limits of the present publication.

R. R.

NEWPARK HOUSE, MID-CALDER,
7th March 1878.

NOTE TO PP. 25 AND 26.

Mid-Lothian Fiars Prices for 1877.

These were struck on 8th March 1878, under the presidency of Sheriff Davidson. They are as follows (calculated per imp. quarter):—Best wheat, 35s. 9d.; second wheat, 33s. 3d.; best barley, 31s. 7d.; second barley, 29s. 6d.; third barley, 28s.; best oats, 27s. 3d.; second oats, 25s.; peas and beans, 43s. 4d.; oatmeal per cwt. of 112 lbs. avoird., 16s. 4d.; oatmeal per load of 280 lbs. avoird., being 2 imp. bolls, 40s. 10d. The following Farmers formed the jury referred to at p. 26 :—*Low district*—John H. Dickson, Saughton Mains, Corstorphine; James Hope, Easter Duddingston; James Stenhouse, South Gyle, Corstorphine; W. Ford, Hardengreen, Dalkeith; Robert T. Harper, Edmonstone Mains, Liberton. *Middle district*—A. G. Cunningham, Rosebank, Currie; James Wilson, Wester Cowden, Dalkeith; John Hunter, Oxenford Mains, Ormiston; P. Brotherston, Sauchenside, Dalkeith. *High district*—George Davidson, Dean Park, Balerno; William Steuart, Selms, Kirknewton; James Hunter, Harrymuir, Mid-Calder.

PRINTED BY NEILL AND CO., EDINBURGH.

APPENDIX by Mr Peter Loney, Edinburgh, 1895.

Late Observer for the Meteorological Societies of Edinburgh and London.

Mean Temperature at Newington (Elevation 270 feet), Edinburgh.

Years	1877	1878	1879	1880	1881	1882	1883	1884	1885	1886	1887	1888	1889	1890	1891	1892	1893	1894	Mean.
January,	39	33	31	37	29	42	39	41	37	35	39	39	40	42	37	37	39	38	37·6
February,	41	42	35	43	36	43	41	41	41	35	40	36	37	38	43	37	40	40	39·4
March,	38	41	37	41	39	44	37	42	40	38	39	36	40	43	38	37	41	44	39·9
April,	42	46	41	40	43	44	46	45	44	43	43	43	43	45	43	44	48	48	44·3
May,	46	51	46	50	51	50	49	49	47	48	49	50	52	51	48	51	54	47	49·1
June,	57	57	52	55	55	54	54	55	55	54	58	52	58	55	55	54	50	55	55·2
July,	58	61	54	58	58	58	56	57	60	58	61	55	58	57	59	56	58	59	57·8
August,	55	59	56	61	55	57	58	59	55	58	58	56	57	58	57	57	62	57	57·4
September,	51	55	52	55	54	51	54	55	52	54	53	52	53	58	56	52	54	52	53·5
October,	47	49	46	44	44	48	48	49	44	50	45	48	46	49	48	44	49	46	46·9
November,	43	38	40	40	46	40	42	42	41	45	40	43	44	42	41	43	41	46	42·2
December,	40	31	35	38	39	34	41	38	39	35	37	41	40	35	40	35	42	41	37·8
Total and Mean Temperatures,	46·4	47·3	43·8	47·3	45·7	47·1	47	47·7	46·2	46	47	46	47·3	47·6	47·1	45·6	49	47·7	46·7

RAINFALL AT NEWINGTON (Elevation 270 feet), EDINBURGH.

YEARS	1877	1878	1879	1880	1881	1882	1883	1884	1885	1886	1887	1888	1889	1890	1891	1892	1893	1894	Average.
	in.	in.	in.	in.	in.	in.	in.	in.	in.	in.	in.	in.	in.	in.	in.	in.	in.	in.	in.
January,	7·42	3·84	1·30	0·77	1·76	1·91	3·04	3·64	1·32	4·38	0·96	1·71	0·64	3·70	0·73	1·05	0·71	2·47	2·30
February,	2·22	0·86	2·05	1·83	4·61	1·79	1·38	1·55	2·26	1·50	1·09	1·45	1·17	0·95	0·21	2·00	2·63	6·81	2·02
March,	2·67	0·78	2·44	1·62	1·84	2·75	1·35	1·80	1·68	2·76	1·70	3·67	0·78	1·83	3·01	1·37	0·83	1·71	1·92
April,	3·46	1·50	2·69	3·70	1·19	2·71	1·64	1·22	2·06	1·53	1·88	1·56	3·09	0·84	0·29	1·11	1·64	1·75	1·88
May,	2·56	3·02	1·79	0·90	1·91	2·24	0·84	2·42	3·08	4·83	1·47	0·76	1·59	1·71	1·91	3·00	1·37	3·20	2·14
June,	1·80	1·81	4·99	1·90	1·79	2·68	1·77	0·96	0·54	2·24	0·35	2·38	1·49	3·23	0·57	3·14	2·48	2·65	2·04
July,	4·76	0·76	7·63	4·47	3·08	4·53	4·21	4·63	0·90	3·01	2·21	5·71	3·96	2·43	3·69	1·17	2·53	2·83	3·47
August,	9·64	4·54	2·73	0·55	6·07	1·57	3·09	2·09	2·77	0·86	1·65	1·88	5·40	4·28	5·21	4·80	3·16	3·83	3·56
September,	1·33	2·93	1·60	2·84	3·26	1·77	1·62	2·36	2·81	2·47	4·78	0·59	0·87	2·23	4·71	1·20	1·67	0·48	2·19
October,	2·72	1·87	1·28	4·15	1·99	2·65	2·00	1·50	1·30	4·45	1·44	1·18	3·33	3·05	2·03	3·92	2·51	2·79	2·46
November,	2·42	3·67	2·42	3·84	2·90	3·54	2·52	1·45	1·27	1·28	3·37	4·77	0·68	4·84	1·69	1·43	1·74	1·43	2·51
December,	1·62	2·81	1·71	3·51	1·67	4·69	1·14	3·63	1·17	2·57	1·23	1·03	1·42	1·55	5·16	1·36	2·50	2·25	2·28
Total,	42·62	28·39	32·63	30·08	32·07	32·83	24·69	27·25	21·16	31·88	22·13	26·69	24·42	30·64	29·21	25·55	23·82	32·20	28·27

The Fiars Prices for the County of Edinburgh from the Years 1877 to 1894.

	1877		1878		1879		1880		1881		1882		1883		1884		1885	
	s.	d.	s.	d.	s.	d.	s.	d.	s.	d.	s.	d.	s.	d.	s.	d.	s.	d.
Wheat, 1st, p. Qr.,	35	9	38	0	37	$10\frac{1}{2}$	42	0	33	4	36	7	35	2	29	9	28	1
,, 2nd, ,,	33	3	36	0	35	6	40	0	31	6	34	6	33	4	28	0	26	7
Barley, 1st, ,,	31	7	35	2	33	7	34	10	26	3	30	7	28	9	28	7	25	7
,, 2nd, ,,	29	6	33	8	32	0	33	0	25	0	29	0	27	0	27	0	24	4
,, 3rd, ,,	28	0	32	6	30	0	31	6	24	0	28	0	25	6	26	0	23	4
Oats, 1st, ,,	27	3	22	1	25	0	22	8	22	10	23	9	24	0	22	7	23	3
,, 2nd, ,,	25	0	20	3	23	0	21	0	21	4	22	0	22	3	21	0	21	6
Peas and Beans,	43	4	36	11	44	0	41	8	37	8	35	5	36	3	32	$7\frac{1}{2}$	30	3
Oatmeal, p. Boll of 140 lbs.,	20	5	16	$6\frac{3}{4}$	18	9	17	0	17	$1\frac{1}{2}$	17	$9\frac{3}{4}$	18	0	16	$11\frac{1}{4}$	17	5

	1886		1887		1888		1889		1890		1891		1892		1893		1894	
	s.	d.	s.	d.	s.	d.	s.	d.	s.	d.	s.	d.	s.	d.	s.	d.	s.	d.
Wheat, 1st, p. Qr.,	26	6	28	8	26	8	29	4	30	6	35	$11\frac{1}{2}$	26	$9\frac{1}{2}$	25	$4\frac{1}{2}$	21	$6\frac{3}{4}$
,, 2nd, ,,	25	0	27	0	25	6	27	10	29	0	34	0	25	3	24	0	20	4
Barley, 1st, ,,	24	0	27	4	26	3	28	7	26	7	30	0	23	$9\frac{3}{4}$	29	$6\frac{1}{2}$	23	3
,, 2nd, ,,	22	6	25	6	24	6	26	9	25	0	28	3	22	3	28	0	22	0
,, 3rd, ,,	21	0	24	0	23	0	25	3	23	9	26	9	21	0	26	6	21	0
Oats, 1st, ,,	19	6	17	10	19	11	21	8	20	0	24	11	21	10	20	$3\frac{1}{2}$	18	8
,, 2nd, ,,	18	0	16	6	18	6	20	0	18	6	23	6	20	4	19	0	17	6
Peas and Beans,	27	0	32	0	40	0	29	$10\frac{1}{2}$	29	10	33	6	29	4	...		23	0
Oatmeal, p. Boll of 140 lbs.,	14	$7\frac{1}{2}$	13	$4\frac{1}{2}$	14	$11\frac{1}{4}$	16	3	15	0	18	$\frac{1}{2}$	16	$4\frac{1}{2}$	15	$2\frac{7}{12}$	14	0

Average Prices for Ten Years from 1877 to 1886 inclusive.

	s.	d.	
Wheat, 1st,	34	$3\frac{1}{2}\frac{6}{10}$	per Quarter.
,, 2nd,	30	$4\frac{1}{4}\frac{6}{10}$	,,
Barley, 1st,	29	$10\frac{1}{2}\frac{8}{10}$	,,
,, 2nd,	28	$3\frac{1}{2}\frac{4}{10}$	,,
,, 3rd,	26	$11\frac{3}{4}\frac{7}{10}$	,,
Oats, 1st,	23	$3\frac{1}{2}$	,,
,, 2nd,	21	$4\frac{1}{4}\frac{9}{10}$	,,
Peas and Beans,	39	$2\frac{1}{2}\frac{2}{10}$	,,
Oatmeal,	17	$5\frac{1}{2}\frac{3}{10}$	per Boll.

Average Prices for 8 Years from 1887 to 1894 inclusive.

	s.	d.	
Wheat, 1st,	28	$1\frac{1}{4}\frac{1}{8}$	per Quarter.
,, 2nd,	26	$7\frac{1}{4}\frac{3}{8}$	,,
Barley, 1st,	26	11	,,
,, 2nd,	25	$3\frac{1}{4}\frac{4}{8}$	,,
,, 3rd,	23	$10\frac{3}{4}\frac{4}{8}$	,,
Oats, 1st,	20	$7\frac{1}{2}\frac{6}{8}$	,,
,, 2nd,	19	$2\frac{3}{4}$	,,
Peas and Beans,	31	$0\frac{3}{4}\frac{5}{7}$	,,
Oatmeal,	15	$5\frac{3}{4}$	per Boll.

In year 1893 Peas and Beans are not quoted in the prices.

EDINBURGH WEATHER NOTES from 1877 to 1894.

The Warmest Years were . .	1884 .	Mean Temperature,	47°·7	
,, ,, . .	1893 .	,,	,,	49°
,, ,, . .	1894 .	,,	,,	47°·7
The Coldest Years were . .	1879 .	,,	,,	43°·8
,, ,, . .	1881 .	,,	,,	45°·7
,, ,, . .	1892 .	,,	,,	45°·6
The Dryest Years were . .	1885 .	Amount of Rainfall,	21·26 inches.	
,, ,, . .	1887 .	,,	,,	22·13 ,,
,, ,, . .	1893 .	,,	,,	23·82 ,,
The Dryest Months, January	1887 .	,,	,,	0·35 ,,
,, ,, February	1891 .	,,	,,	0·21 ,,
,, ,, April .	1891 .	,,	,,	0·29 ,,
The Wettest Years were . .	1877 .	,,	,,	42·62 ,,
,, ,, . .	1882 .	,,	,,	32·83 ,,
,, ,, . .	1894 .	,,	,,	32·20 ,,
The Wettest Months, January	1877 .	,,	,,	7·42 ,,
,, ,, August	1877 .	,,	,,	9·64 ,,
,, ,, July .	1879 .	,,	,,	7·63 ,,

PETER LONEY.

POPULAR WORKS ON NATURAL HISTORY.

Familiar Wild Flowers, described by SHIRLEY HIBBERD, and *illustrated with 200 ... All with COLOURED PLATES, taken from Nature, from original paintings by ... E. Hulme, F.L.S., &c., and numerous wood engravings, 5 vols., post 8vo, cloth, gilt tops, new, 42s. (published £3, 2s. 6d.)*

Familiar Garden Flowers, by the same Editor, and *illustrated with 200 coloured PLATES, and uniform with above, 5 vols., 42s. (published £3, 2s. 6d.)*

Familiar Wild Birds, with descriptive text, by W. SWAYSLAND, *illustrated with 160 charming full-page COLOURED PLATES, from Nature, and numerous original engravings, 4 vols., uniform with above, new, 30s. (published 50s.)*

Familiar Trees, with descriptive text by G. S. BOULGER, F.L.S., &c., complete set, *illustrated with ... full-page beautifully COLOURED PLATES and woodcuts by W. H. J. BOOT, 2 vols., uniform with above, 12s. (published 25s.)*

A charming work, useful not only to the botanist, but of considerable interest to the artist and the poet, and also, from their many associations to the historian and moralist, the student of literature and of folk-lore. Complete sets of these beautifully illustrated works rarely occur for sale at second-hand prices.

NEW EDITIONS of the Rev. F. O. MORRIS'S WORKS.

A History of British Birds. THIRD EDITION, newly revised, corrected, and enlarged by the AUTHOR, *illustrated with 394 beautiful HAND-COLOURED PLATES*, with the letter thoroughly revised and enlarged, 6 vols., super royal 8vo, cloth, new, £4, 10s. (published £6, 6s.)

"It is a work which every lover of nature, everyone who wishes to become intimately acquainted with the feathered tribes of our land, ought by all means to possess himself of. The extremely pleasing and lively manner in which every bird is introduced, its history given, it habits and peculiarities described, and the numerous anecdotes contained in each history related, impart a charm of no ordinary kind to the work, a charm only equalled by that which pervades the 'History of British Butterflies' by the same author."—*Naturalist.*

―――― **The Same.** *New Issue in Thirty-six Monthly Parts, with the 394 Plates all coloured by hand, price 1s. 6d. per part.*

A Natural History of the Nests and Eggs of British Birds. THIRD EDITION, newly revised, corrected and enlarged by the AUTHOR, *illustrated with 248 COLOURED PLATES, 3 vols., super royal 8vo, cloth, new, £2, 5s. (published £3, 3s.)*

Intended as a supplement to the "History of British Birds," this work gives the fullest information respecting the localities and construction of their nests, the number and peculiarities of their eggs, and all the instruction requisite for determining to what species they belong.

"The markings of the eggs, many of which are figured in their natural size, are particularly true and distinct, and the plates must of but prove exceedingly valuable to collectors. All the nests and eggs have been drawn from actual specimens, in most cases the names of the persons from whom they either obtained the objects being given."

A Natural History of British Moths. THIRD EDITION, *with 132 plates, containing nearly 2000 beautifully COLOURED specimens, 4 vols., royal 8vo, cloth, new, £3, 3s. (published £6, 6s.)*

Accurately delineating every known species, with the English as well as the scientific names, accompanied by full description, date of appearance, list of the localities they haunt, their food in the caterpillar state, and other features of their habits and modes of existence, etc.

"Speaking of entomology, we should place Mr. Morris's 'History of British Moths' at the head. It gives coloured figure of every known British moth, together with dates of appearance, localities, description, and food of caterpillar; is form a handsome work for a library, and will, we should hope, lead many to commence the fascinating study of entomology.

A History of British Butterflies. SEVENTH EDITION, *newly revised, corrected, and enlarged by the AUTHOR, illustrated with 71 BEAUTIFULLY COLOURED plates, super royal 8vo, cloth, new, 15s. (published £1, 1s.)*

"With coloured illustrations of all the species, and separate figures of the male and female where there is any obvious difference between them, and also of the under side, together with the Caterpillar and Chrysalis. Add a full description of each, with copious accounts of their several habits, localities, and dates of appearance, together with details as to their preservation, and, with new and valuable information, the result of the author's experience for many years."

―――― **The Same.** *New Issue in Six Monthly Parts, with the 71 Plates all coloured by hand, price 2s. 6d. per part.*

Ferns, British and Exotic, by E. J. LOWE, F.R.S., etc. *Illustrated with 479 ... COLOURED plates, beautifully COLOURED FROM NATURE, and numerous wood engravings, 8 vols., royal 8vo, cloth, as new, £4, 10s. (published £6, 6s.)*

The standard popular work on this graceful decorative plant, giving, along with an exact drawing of each species, its classification and nomenclature, and pointing out the minute differences which in many instances form the chief distinction between the species.

Lightning Source UK Ltd.
Milton Keynes UK
UKHW050623031022
409835UK00008B/895